DEFENDERS OF THE FUTURE

TACKLE TODAY'S WATER TROUBLES

A Story with Lessons, Activities & Resources to Solve Real-World Problems for Grades 4+

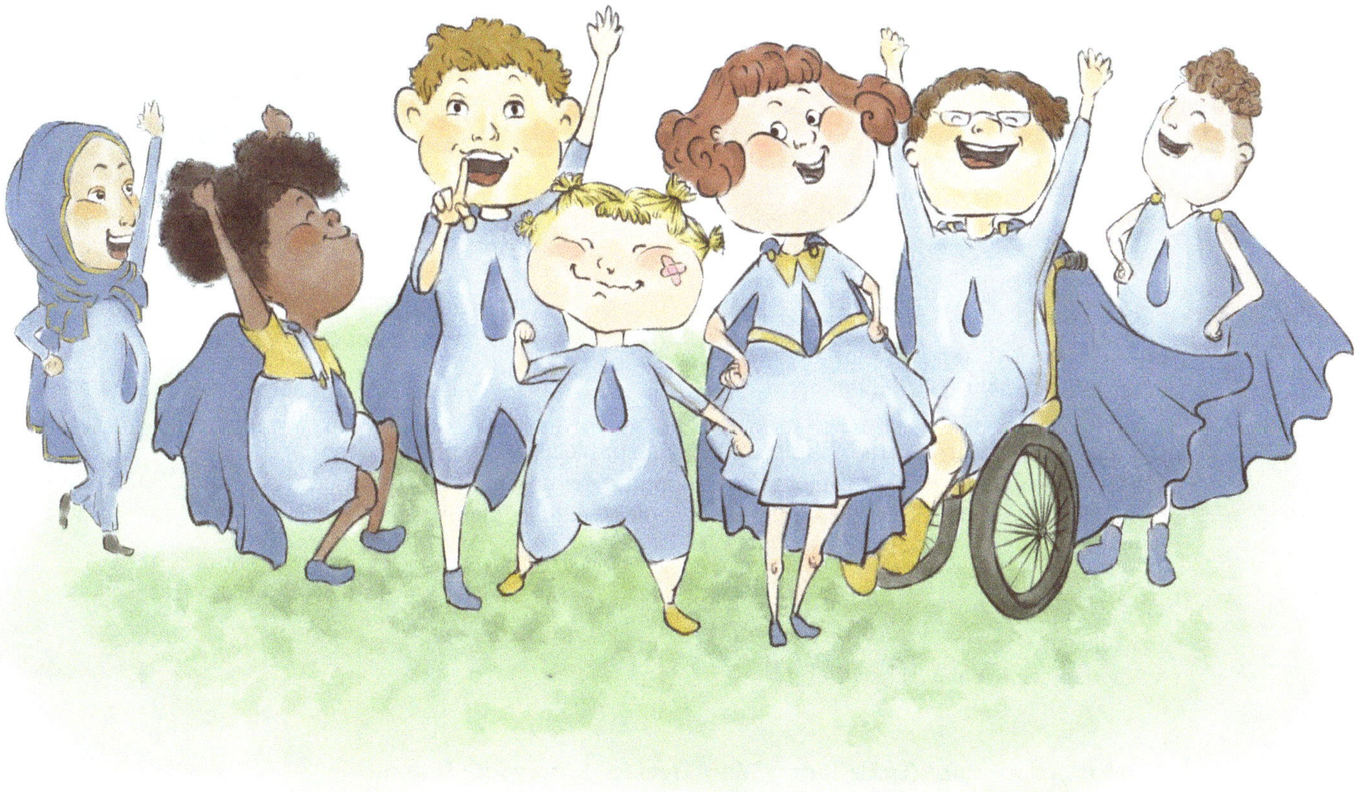

WRITTEN BY **DAWN PAPE**
ILLUSTRATED BY **LISA SACCHI**

GOOD GREEN LIFE
— publishing —

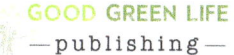

Text copyright © 2020 Dawn Pape
Illustration copyright © 2020 Lisa Sacchi

All rights reserved, but all parts of this book may be reproduced with consent of the publisher. Content may be used without permission in the case of brief excerpts used in critical reviews and articles. For longer excerpts, please address inquiries to:

Good Green Life Publishing
P.O. Box 74
Circle Pines, MN 55414
www.goodgreenlifepublishing.com

Pape, Dawn Viola.
Defenders of the Future—Tackle Today's Water Troubles. A Story with Lessons, Activities, and Resources for Grades 4+ / written by Dawn Pape, illustrated by Lisa Sacchi
Summary: The water cycle character explains how humans are impacting water and what can be done to protect water resources.

ISBN 978-0-9971131-8-1
[1. water cycle education 2. stormwater runoff education 3. empower children to solve problems 4. environmental education 5. climate crisis education]

Library of Congress Control Number: 2019903336

The hope for tomorrow rests on the actions we take today.
Let's work together and use our powers for good.

- Dawn Pape

ABOUT

This book doesn't dance around real-world water challenges, but tackles them. Combining science and social studies, the defenders of the future characters interact with the talking water character. The story reveals how, even though humans are negatively impacting water, regular people (like your students!) can use their super strengths to turn these challenges around.

HOW TO USE THIS BOOK

The story can be read aloud by a teacher or performed as a skit. Since this story is filled with dense content and big issues, the author suggests following the book's layout and reading only a few pages of the story at a time followed by the corresponding activities (delineated by dotted borders) to help the students absorb the information before moving on to the next section or "mission."

CONTENTS

Foreword
    Discovering Kids' Super Powers to be Defenders of the Future.............................6-11

Introduction..12-27
    History of the Earth..12-13
        Activity 1: Earth Timeline:
            Getting Your Arms Around the Earth's History.................14-15
    The Water Cycle..16-17
        Activity 2: Water Appreciation/Recreation Field Trip 18-19
        Activity 3: Tapping into Super Strengths to
            Demonstrate Water Cycle Knowledge.................................20-21
        Activity 4: Our Planet's Freshwater Availability......................................22-23
    Discovering the "Leaks" in the Water Cycle...24-27

Mission 1: Let Water Soak into the Earth..28-35

Mission 2: Keep Streets Clean..36-37
        Activity 5: What is a Watershed?...38-39
        Activity 6: We All Have Lakeshore Property:
            Stormwater Runoff Travels from Streets to Streams................40-42
        Activity 7: You're in Charge!
            Using "Best Management Practices" (BMPs).........................43-45
    Activity 8: How is U.S. Freshwater Used?..46-49

Mission 3: Use Less Energy..50-55
        Activity 9: Home Energy Use...56-57

Mission 4: Don't Waste Food...58-63
        Activity 10: How Much Water Do You Eat?...64-67

Resources..68-70

Glossary..71-73

FOREWORD

DISCOVERING KIDS' SUPER POWERS TO BE DEFENDERS OF THE FUTURE

Summary

Inspire kids to find and use their strengths to help protect the planet. Have them decide individually, as a group, or as a class how they will help protect water to earn their superhero capes and masks. Explain that even though there is government that oversees environmental resources, government cannot be everywhere all the time enforcing rules and best practices, so it's important for individuals to do their part—even when no one is watching.

Background

Kids tend to be creative and energetic. If we can harness this creative energy and teach kids to find and use their strengths, we can help them develop their self-esteem by knowing they are positively helping to protect natural resources. In addition, teaching kids how to use their strengths to give back to the community could be the start of a life-long, civic-minded lifestyle. Common sense is backed up by research that shows those who volunteer as kids are more likely to continue into adulthood. Dr. Tom Harrison, from the University of Birmingham co-authored and published a report by the Jubilee Centre for Character and Virtues. This report stressed the practical importance of the research for social action providers across the country. He stated, "These findings will help those in the voluntary sector plan and deliver youth social action programmes that support young people to cultivate a habit of service. The more people who contribute to the common good, the more likely we are to flourish as a nation."

> *"Remember, you can have a positive, lifelong effect by helping your child take advantage of their natural patterns of thought, feeling and behavior and showing them how important and valuable their unique talents really are."*
>
> —*StrengthsExplorer.com*

In addition, we've all heard that "it's better to give than to receive." This saying has ancient roots in many religious texts; the Bible, the Torah, and the Quran all encourage helping others. To a kid, this teaching might sound like an excuse for a mediocre present. However, research backs up this seemingly trite statement. It *truly* is better for ones' health to give than to receive. Research consistently shows that doing a good deed such as volunteering is good for us. It has been correlated with better health (lower blood pressure, decreased stress and depression), longer lives, higher self-esteem, long-term happiness, and a greater sense of purpose. Giving back connects us to the community and makes us realize that we are not isolated, yet we are also not the center of the universe. We are part of a much larger whole and common good.

In the Minnesota Twin Cities Metro Area, top threats to water include: excess salt use on sidewalks, streets and parking lots; excess nutrients found in soil, leaves and grass clippings that reach our waters via stormwater runoff; and invasive species. Important volunteer actions could be as simple as raking leaves in the fall, sweeping up excess salt in the winter, adopting a stormdrain (adopt-a-drain.org), or making a Pledge to Plant for Pollinators and Clean Water (bluethumb.org).

Materials

Consider making superhero attire that reflects students' "super powers." In these pictures, blue capes and masks were purchased and the kids decorated them to reflect their talents.

FOREWORD

Identifying Strengths

The first step in helping students on this journey to be "defenders of the future" is to help them identify their strengths or "super powers." When people use their strengths, pitching in on projects is fun and won't feel like work because it involves doing things the students enjoy and are already good at.

There are a number of models to help identify "multiple intelligences" (Howard Gardner) or strengths. For adults, a popular method is using an online Strengths Finder exam. In a similar vein, gallupstrengthcenter.com came up with questions for kids ages 10 – 14 years old to help them discover their passions and develop their talents. There are free downloadable workbooks to help young people discover their strengths with a teacher's guide to accompany it. A more in-depth (and costly) alternative is to have the students take the online quiz.

What are your top three strengths?

1. **Achieving**—like to accomplish things and have a great deal of energy
2. **Caring**—enjoy helping others
3. **Competing**—enjoy measuring their performance against others and have a great desire to win
4. **Confidence**—believe in themselves and their ability to be successful in their endeavors
5. **Dependability**—keep their promises and show a high level of responsibility
6. **Discoverer**—tend to be very curious and like to ask "Why?" and "How?"
7. **Future Thinker**—think about what's possible beyond the present time, even beyond their lifetime
8. **Organizer**—good at scheduling, planning, and organizing
9. **Presence**—like to tell stories and be at the center of attention
10. **Relating**—good at establishing meaningful friendships and maintaining them

Based on Howard Gardner's Multiple Intelligences

Linguistic ("word smart")
finds the right words

Bodily-Kinesthetic ("body smart")
coordinates mind and body

Existential ("cosmic/spiritual smart")
tackles big questions such as why we live and die
Note: Gardner's Theory doesn't include this one

Visual ("spatial/picture smart")
creates and visualizes the world in 3-D

Interpersonal ("people smart")
senses people's feelings, works well with others

Logical-mathematical ("number/reasoning smart")
quantifies, makes hypothesis, strategizes

Intrapersonal ("self smart")
excels at understanding one's own feeling, needs and wants

Musical ("music smart")
discerns and reproduces sounds, pitch, rhythm

Naturalist ("nature smart")
feels connected to living things, can "read" nature

FOREWORD

Project Ideas

How are kids going to commit to being protecting water even when no one is looking? Environmental psychology research in fostering sustainable behavior indicates that if a person publicly commits to making a positive step of caring for the environment, he/she will be more likely to follow-through with the behavior. A fun way to have students make their intentions public is to have the students create a super power costume (mask and cape) that exemplifies their super strengths and how they plan to help protect water. For example, if a student identifies him/herself as a caring, "body-smart" kid, perhaps this cape would be red with hearts on it and a big broom sweeping up streets and sidewalks wherever there is pollution. To keep costs down, check around town to see if you can get fabric donated. Otherwise, masks can be purchased online for less than a dollar each.

Projects take time, so it is a good idea to phase the project over a longer period of time and identify smaller milestones. This project is also a way to find out what students still want to learn and encourage them to keep learning and finding answers. The project ideas listed on the next page are certainly not exhaustive, so encourage your students to be creative and find a meaningful project to do. In addition, a list of more than 200 citizen-science projects for students and adults, with a toolkit for people looking to design and maintain their own local projects is at **CitizenScience.gov**.

Project Ideas Aligned to Multiple Intelligences

Linguistic	Learn how and then teach others how to drain and inspect a watercraft	Write letters to organizations encouraging BMPs to be implemented.	Make and hand out flyers informing businesses about the correct amount of salt to apply (see saltsmart.info)
Visual	Learn how to drain and inspect a boat	Design a raingarden planting plan	Figure out where stormwater is coming from to determine raingarden location
Kinesthetic	Stencil stormdrains with "drains to creek/lake/river" message	**Summer** sweep up grass clippings **Fall** rake leaves, **Winter/Spring** sweep up excess salt	Help build and/or maintain a raingarden
Musical	Teach others about water by performing the Water Cycle	Create a song, poem, or artwork showing about an aspect of water that resonated with the student	Create visuals to support other groups' projects, i.e. handouts, posters, maps...
Interpersonal	Find a raingarden to visit and interview the people who created it	Get people with various intelligences to work together on a project	Encourage others to implement a best management practice, e.g. sweep up excess salt
Intrapersonal	Reduce your consumption ([Video about the problem of plastics](#))	Pledge to... -plant for pollinators and clean water -reduce your planet impact	Eat locally and/or lower on the food chain
Naturalist	Observe nature and describe how it makes you feel	Label the plants and trees at your school	Plant or tend a garden
Logical	Identify largest pollution concerns near your school and determine BMPs to use	Determine raingarden size to handle the volume of stormwater runoff. Also calculate how many plants will need to be planted	Quantify: - your food miles - carbon footprint - water footprint
Existential	What is the best use of water?	Which systems need to change to draw down carbon emissions? ([drawdown.org](#))	What is our responsibility to future generations?

INTRODUCTION

Narrator	One day, a group of kids gathered in the park. They were chatting with each other trying to figure out what to play, when all of a sudden a puddle of water started *talking* to them.
Water	The puddle said quietly, "Hello, my children. I am Water."
Narrator	The kids froze not knowing whether to run or to listen.
	The puddle continued, "I am very, very, very, very, VERY old. I was created a 3.5 billion years ago—just a billion years after the Earth was formed.
	The kids were **mesmerized** by the talking water and realized that, while it was certainly out of the ordinary, the talking water seemed peaceful. They listened intently.
Water	Water continued, "I take credit for giving life to this planet and I've been watching it evolve ever since. I watched as the first bacteria were formed when I was only a half billion years old. I saw how the atmosphere became oxygen-rich. And not too long ago, I observed the dinosaurs come and go. Now people are large and in charge as the dominant species. But from my perspective, you've only been here for a split second."

ACTIVITY 1 — 5-10 Minutes

INTRODUCTION

GETTING OUR ARMS AROUND THE EARTH'S HISTORY

Summary and Purpose
Have students experience the Earth's evolution in a concrete way.

Background
It is important for people to put evolution into perspective to understand the magnitude of people's impact on the planet in a very short amount of time. We are currently in the beginning of the sixth mass extinction in history and many scientists are naming our current geologic era "**Anthropocene.**"(*anthropo* is Greek for "human," and "*-cene*" denotes an epoch of geologic time.

Strengths Used

Materials
Handouts of the Earth's timeline (on the next page) or display the timeline on a screen for students to see.

Procedures
Have students stand up with their arms spread wide and connect when major occurrences happened using their armspan as the timeline. For example, the Earth was formed at the students' left finger tips, the oxygen-rich atmosphere was formed at the students' lungs, dinosaurs lived at the time of students' right hand, and people have been around since the end of students' right finger nails.

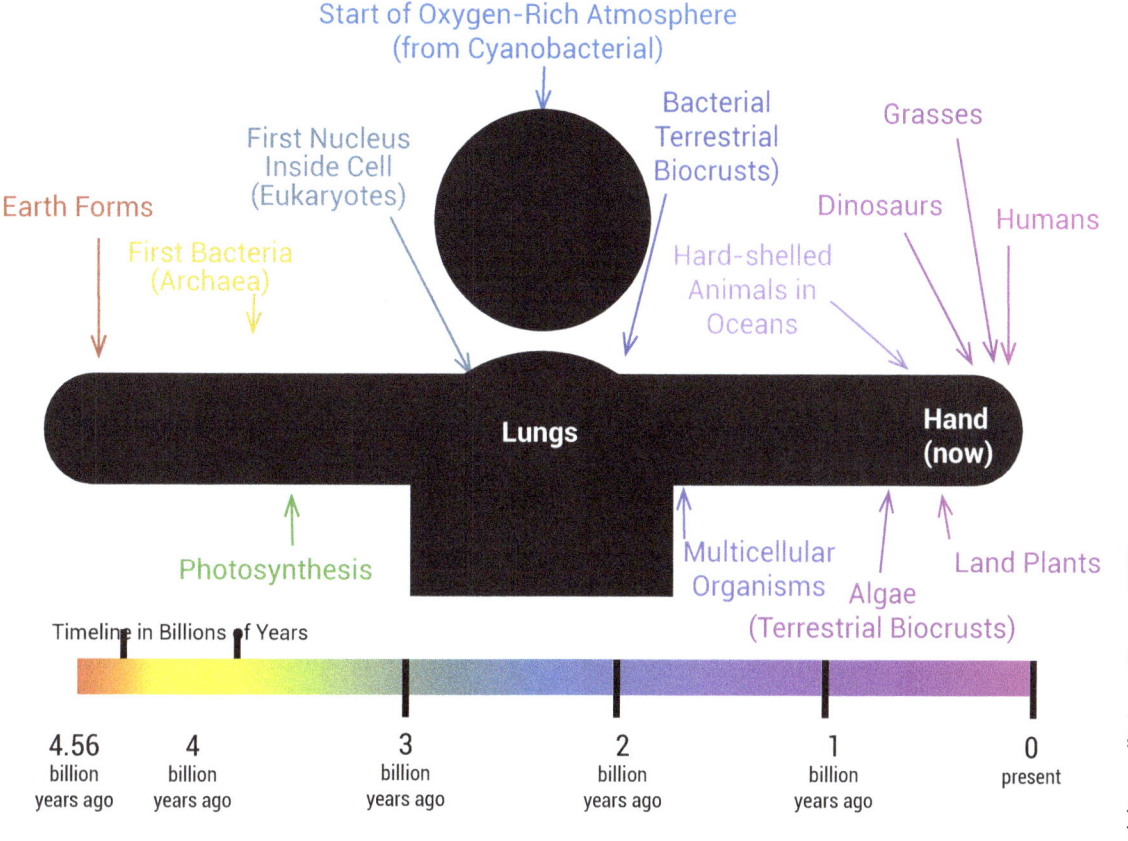

INTRODUCTION CONTINUED

Narrator — The kids were shocked that Water considered people newcomers on the planet. They looked at each other in surprise, but no one spoke. They wanted to hear more from Water.

Water — Water continued talking, "Something you might want to know about me is that I am always on the move above and below the Earth. My movement is called the water—or hydrologic—cycle. There is no beginning or end to the water cycle. I just keep going around and around.

Sometimes my journey is quick and sometimes my cycle can take millions of years to complete. Here's one example of how the water cycle works: the sun evaporates me from lakes, rivers, and oceans—or even this little puddle—and I float up to sky as vapor. As I literally collect myself, I condense and form clouds. Then I travel back down to Earth as precipitation, such as rain, sleet, or snow. Finally, I often collect in a lake or river. Or, the I might fall on then land and soak in—or infiltrate—into the ground.

Precipitation, Infiltration, Collection, Evaporation, Condensation.

I was quite happy doing my water thing for billions and billions and billions of years. **Precipitation, infiltration, collection, evaporation, condensation**. Repeat. Precipitation, infiltration, collection, evaporation, condensation. Repeat.

I felt appreciated by the Earth's inhabitants and even respected. It was a fun and peaceful time for me."

ACTIVITY 2  Variable

INTRODUCTION

WATER APPRECIATION/ RECREATION FIELD TRIP

Summary and Purpose

Use water for recreation—canoeing, kayaking, paddle boating, pontoon ride, swimming, fishing, ice skating... Developing a relationship and an affinity for water lays the groundwork for wanting to protect it. If time or budgets are constraints consider a simple activity like running through a sprinkler or a water balloon toss game.

Strengths Used

ACTIVITY 3
15-45 Minutes

INTRODUCTION

TAPPING INTO SUPER STRENGTHS TO DEMONSTRATE WATER CYCLE KNOWLEDGE

Summary and Purpose
Students recognize their talents and demonstrate their knowledge of the water cycle using their strengths.

Background
Water makes up around 70% of all living matter and is one of the most important components of ecosystems. Water cycle's three major processes: evaporation, condensation, precipitation involve the exchange of energy and temperature changes. When water evaporates, it takes up energy from its surroundings and cools the environment. When it condenses, it releases energy and warms the environment. These factors are the basis of understanding heat exchanges which influence climate.

Strengths Used

Materials
Materials will vary depending on the strengths students choose. Supplies may include paper, markers, pencils, a YouTube video (for example, to listen to a meditation on water), and/or musical instruments

Procedures
Have students identify their own strengths by walking the kids through various types of intelligences listed on the following page. If students have a hard time identifying their strengths, ask them which way sound like the most fun. If something is fun, it is likely a strength of theirs. Then have students use one of their strengths to demonstrate their knowledge of the water cycle. You may want to have students display or perform their projects to the class.

PROJECT IDEAS

Multiple Intelligences
based on Howard Gardner's Theory of Multiple Intelligences

- **INTRAPERSONAL** — *excels at understanding one's own feeling, needs and wants*
  - What does water cycle means to you? How you interact with the water cycle?

- **NATURALIST** — *feels connected to living things, can "read" nature*
  - Teach how to fish or how to "read" at stream.

- **MUSICAL** — *discerns and reproduces sounds, pitch, rhythm*
  - What does the water cycle sound like?

- **LINGUISTIC** — *finds the right words*
  - Describe the different paths water can take in the water cycle.

- **KINESTHETIC** — *coordinates mind and body*
  - Do the water cycle as a dance and/or dramatization.

- **EXISTENTIAL** — *tackles big questions such as why we live and die*
  - Meditate on the process of the water cycle (search "YouTube Water Cycle Meditation").

- **VISUAL** — *also called "spatial," creates and visualizes the world in 3-D*
  - Draw the water cycle.

- **INTERPERSONAL** — *senses people's feelings, works well with others*
  - Collaborate with others to do a larger project.

- **LOGICAL** — *quantifies, makes hypothesis, strategizes*
  - How much water does your household use?

ACTIVITY 4 — 45-60 Minutes

INTRODUCTION

OUR PLANET'S FRESHWATER AVAILABILITY

Summary and Purpose

Since Earth is called "the water planet," why is only 1% available to drink? How do we use our freshwater? If you were in charge, would you make the same freshwater use choices? Kids know that water is necessary for all life and that there is a lot of water on the planet. This exercise can be done as a demonstration to the whole class by the teacher or it can be done in small groups if you have more time and want the kids to practice estimating fractions.

After finding out what a scarce commodity freshwater is, students will analyze how freshwater is used across the country and form opinions about what they consider wise water use to look like. Would they do things differently if they were in charge?

To take this lesson a step further and bring in social studies, learn about the governmental units (such as city, county, watershed district or management organization, and/or soil and water conservation district) that oversee water resources in their region.

Strengths Used

Materials
- Globe or world map
- Butter knife (or knives)
- 1 apple for teacher demonstration or 1 apple per group
- Copies of handout on page 27
- Pencils for students

Procedures

If the World Were An Apple, How Much Fresh Water Would We Have?

(Adapted from Junior Master Gardener Teacher & Leader Guide Level 1, p. 35)

Pretend the Earth is an apple. Cut the apple into quarters to represent the land and water on the planet. After looking at the globe, can students estimate the how many of those sections are water and how many are land? (Approximately 3/4 is water and 1/4 is land.)

Set the "land" quarter aside and slice off <u>**1/8 from one of the quarters**</u>. This thin slice represents all of the <u>**freshwater**</u> that is on the planet—about 3%.

Next, cut this 1/8 slice into thirds. Two-thirds are unavailable to drink because they represent the freshwater that is frozen in glaciers and polar ice caps. The remaining 1/3—about the size of a pea—represents the amount of freshwater available for us to drink.

After seeing this representation of the amount of freshwater available, ask the kids how important it is to conserve this water on a scale of 1-10.

INTRODUCTION

Water	"But recently..." said Water.
Kid 1	"I respect water!" blurted the big kid. "In fact, I LOVE it. Swimming is my favorite thing to do!"
Water	"That's nice! But..." Water tried to continue.
Kid 2	"Me too!" exclaimed the next kid. "I especially love water slides!"
Kid 3	"And water blasters!" hollered another kid in the back as he pretended to squirt his friends.
Water	Water replied, "I'm so glad you all enjoy water so much. Water is fun! Water is also necessary for all life. And..."
Kid 4	"I tried paddle boarding before," chimed in a girl oblivious to the fact that she, too, had just interrupted Water.
Kid 5	The next kid declared with her finger in the air, "My mom said I need to drink half my weight in ounces every day to stay healthy. I weigh 68 pounds so that means I need to drink 34 ounces of water every day."
Narrator	The rest of the kids started telling each other how much water they should drink and excitedly shared their water stories with each other.
	After awhile, the kids' chatter subsided and their attention returned to water. With eyes closed, Water took deep breaths in and exhaled very s-l-o-w-l-y. Soon the whole group was breathing in unison. Water had such a calming effect on them.
Water	Water began again quietly, "Just as each of you have interrupted my story, unfortunately, so are humans interrupting the water cycle. Pollution is a problem too."

Kid 6	**"Oh no! We need water!"** exclaimed a boy.
All Kids	"Is there anything we can do, Water?" cried the concerned kids.
Water	Water replied quickly, "Of course there are things we can do! Where there's a problem, there's always a solution. In this case, there are MANY solutions. And since everyone needs water, we need everyone's help to protect it."
All Kids	**"We'll help, Water!"** cried the kids.

INTRODUCTION continued

Narrator — Suddenly, the kids' clothes changed into superhero attire. With wide eyes and mouths hanging open, they stared at each others' cool outfits.

MISSION 1

Kid 7 — "We...are...superheroes now?" uttered one kid quietly.

Water — "Yes," said Water. "From now on, you will know what needs to be done. You are defenders of the future. You don't even need to wear these clothes to do your missions. You will know that you have the power to change the world no matter what clothes you are wearing! Kids are powerful with creative ideas and much more energy than adults. Your job is to use your energy for good and to make good choices for water because it's the most important resource on the planet and it desperately needs protection."

All Kids   **"Give us our missions, Water!"** cheered the kids, "We are ready!"

MISSION I CONTINUED

Water   Water announced, **"MISSION ONE: LET WATER SOAK INTO THE EARTH** to filter out pollutants and **recharge** the groundwater."

Kid 5   "Why isn't the water soaking into the ground, Water?" asked a thoughtful girl.

Water   "Because our hard people spaces—like roads, roofs, sidewalks and parking lots get in the way. The water can't soak in. It just runs off these **impervious** surfaces and carries the pollutants in its path to the nearest lake or river," said Water.

Kid 1   "Well, Water," argued another, "We need our people spaces so I don't know how we can solve this."

MISSION 1 CONTINUED

Water

"Oh! There are lots of ways," answered Water, "The secret is to mimic my natural water cycle. For example, plant **raingardens** to catch the dirty runoff and let the water soak into the ground. Swales and buffers along farm fields also do an enormous amount of good at capturing the runoff."

RAIN GARDEN

Kid 7	"Oh, that's interesting," remarked the quiet one, "So we can have our people places AND protect water!"
Water	"Yes," answered Water. "There are also different types of **pervious** roads, parking lots, driveways and sidewalks that allow the water to soak in as it would in nature."

MISSION 1

Narrator
Soon the kids were flying above the neighborhood—higher and higher they went. The couldn't believe they could fly! They could see problem spots. Soil from the Johnson farm was running off into the creek.

Image credit: The Watershed Game, University of Minnesota

The parking lot at Garcia's Groceries was draining to a lake.

They decided to divide up the **watershed** and figure out the problem spots so they could implement best management practices to absorb the excess stormwater runoff. They made a plan to tackle all of the problem spots.

Satisfied with their understanding and problem solving. Water bobbed happily. It was definitely time for the second mission.

MISSION 2

Water

Water wasted no time and declared, "**Your SECOND MISSION IS: ALWAYS KEEP YOUR STREETS CLEAN.** Whatever is on your street ends up in our lakes and creeks, because streets are connected to nearby waters with underground pipes. Be especially mindful of sweeping up soil, leaves and grass clippings during the growing season and salt in cold and icy conditions."

All Kids **"WE'RE ON IT!"** shouted the enthusiastic kids. The defenders went right to work and raked up leaves, and swept up soil and grass clippings.

Narrator Happy with their progress, Water brought the group together and said,

Water "OK, friends, I see you are ready for the next mission." The kids nodded their heads eagerly.

ACTIVITY 5
45-60 Minutes

MISSION 2

WHAT IS A WATERSHED?

Summary and Purpose

This lesson combines science and social studies by introducing the concepts of watersheds and special-purpose local units of government that oversee water resources within a watershed.

In exploring how water moves across the land, students will also be introduced to the concept of stormwater runoff as the students go outside and see how the water drains from their school grounds. They can then go to the USGS website (https://water.usgs.gov/wsc/map_index.html) to zoom in on their watershed and see how they connect to larger watersheds.

Depending on what information is collected in your region and shared with the USGS, you can examine everything from real-time stream flows, water quality samples, groundwater inventory levels, precipitation, drought and flood conditions, water temperature, U.S. water use, current news and developing issues affecting watersheds, aquatic invasive (or non-indigenous) species, watershed education, water pollution report about herbicides, pharmaceuticals, hormones, and other organic wastewater contaminants in U.S. streams, and much more.

Background: What's a Watershed?

A watershed is an area of land that drains to a particular water body. Watersheds can be subdivided into smaller and smaller watersheds. For example, the Mississippi River Watershed drains about two-thirds of the continental U.S. and is divided into six main sub-watersheds. The Upper Mississippi River Basin is made up of many other smaller sub-watersheds.

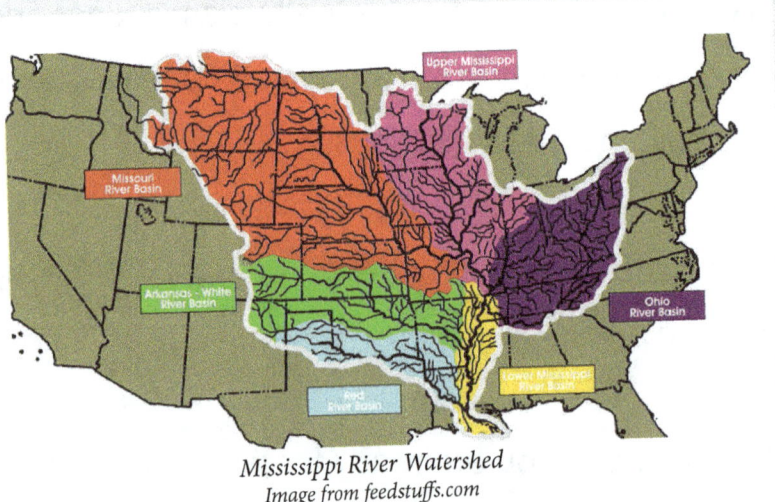
Mississippi River Watershed
Image from feedstuffs.com

Strengths Used

Materials

Outside
- Sidewalk chalk
- Bucket or jug of water
- Computers or smart board

Inside
- Paper
- Markers
- Spray bottles
- Computers or smart board

Procedures

OUTDOORS: WHERE DOES THE WATER FLOW? Go outside with the students with sidewalk chalk and a couple gallons of water. Ask students to guess which way the water will flow from various places on the school grounds. Have them mark their guesses with an arrow pointing in the direction they think the water will flow. Test their guesses by pouring water in the area. Introduce the concept of a watershed—an area of land that drains to a particular body of water. Ask students if they know where the water from the school grounds flows next. Go inside and visit the USGS website (https://water.usgs.gov/wsc/map_index.html) to learn more about watersheds.

INDOOR ALTERNATIVE: CREATE A MINI-WATERSHED? Crumple up a piece of paper. Draw or scribble on it with markers to represent, soil or "pollution." Lightly spray the paper with water and watch the "pollution" drain to the lowest point. Were they able to predict which way the water would drain?

Where does our school's water go?

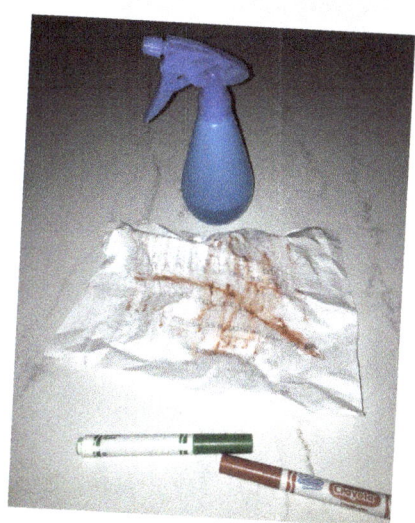

Paper Watershed Model

ACTIVITY 6  20-30 Minutes

MISSION 2

WE ALL HAVE LAKESHORE PROPERTY: STORMWATER RUNOFF TRAVELS FROM STREETS TO STREAMS

Summary and Purpose
Discover how land use affects the water cycle—especially how impervious surfaces don't allow for groundwater recharge.

Background
If the students aren't familiar with the concept of **stormwater runoff**, this video may be helpful: 30-second YouTube "Rubber Ducky Video" (https://www.youtube.com/watch?v=xRKqxOvWx5c) by WaterShed Partners: Minnesota Water–Let's Keep It Clean Campaign, narrated by Ron Schara.

Strengths

Materials
- Plastic paint tray liners
- Glad Press 'n Seal-type wrap
- Permanent markers
- Dry Jell-o chocolate pudding powder
- Dry green Jell-o chocolate pudding powder
- Dry red Jell-o
- Colored and/or chocolate sprinkles (other types of pollution, pet waste)
- Spray bottles with water
- Sink or empty bucket (to pour watershed contents into)

Procedures

Create Your Own Best Management Practice (BMP)

Give each group a paint tray liner and inform the students that the depression at one end of the paint tray is the "lake" and their job is to keep the lake clean. Have students create a watershed including places to work, live, recreate and shop. HINT: if you want to reuse the paint trays, have the students draw on Glad Press 'n Seal-type wrap to put over the paint trays.

After the students have drawn their watershed with permanent markers, have the students imagine what type(s) of pollution their land uses might be causing. At this point, the teacher (or an extremely careful student helper) goes around to add "pollutants," aka a tiny pinch of dry J-ello powder. Chocolate pudding powder makes good "soil," green J-ello can be "grass clippings," and chocolate sprinkles can be imaginary "pet waste." The next step is to make it "rain" with a couple sprays from the spray bottle.

The next step is to add BMPs to stop the pollution. Pieces of felt can be used as "grass swales" or "silt fences." Sponges of varying thicknesses cut to different sizes can be pretend "native plants, raingardens and/or wetlands." Straws can represent pipes and/or draintile, etc. Students can use existing BMPs they learned about in the Watershed Game or they can make up their own practices and techniques. Perhaps students will realize that their watershed was improperly designed and a lot of pollution could be reduced with better city planning.

STEAM Approach

Identify the problem
Research the problem
Develop possible solutions
Select the best solution
Construct a prototype
Test and evaluate
Communicate the solution
Redesign as necessary

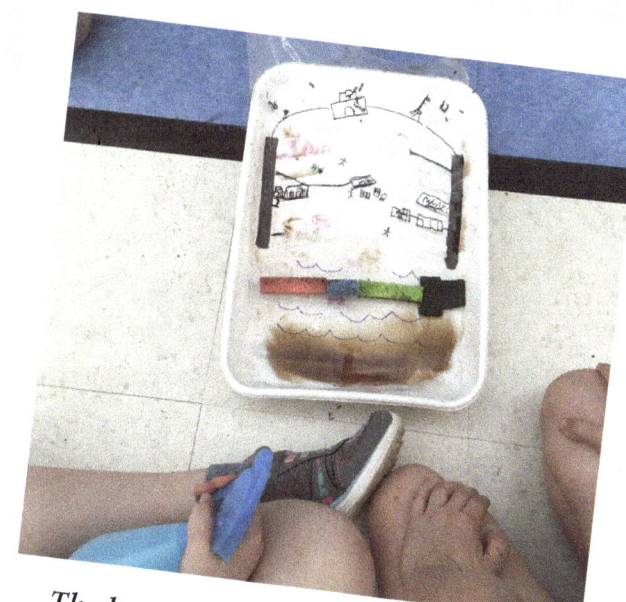

The beauty of this BMP model is that students can empty their tray and try again.

Option for Active Groups: Be a Stormdrain Goalie

Clear desks aside or go to a larger space. Someone rolls the ping-pong balls attempting to make it between the cones that represents a **stormdrain**. Other students stop "pollution" (ping-pong balls) from going into the "stormdrain" (area between two cones). Different colored balls can represent different types of pollution.

Stormdrain Goalie

ACTIVITY 7
45-60 Minutes

MISSION 2

YOU'RE IN CHARGE!
USING "BEST MANAGEMENT PRACTICES" (BMPS) IN YOUR COMMUNITY

Summary and Purpose
This activity allows students to role play as various members of the community and to work together to learn about and solve water pollution problems by implementing various tools. Students discover that land use teams must work together as a watershed to meet the Watershed Clean Water Goal. By the end of the game, students will be able to describe how human activities associated with various land uses within a watershed may pollute a stream with excess sediment or phosphorus. They will also understand how excess sediment and phosphorus can affect the health of a stream by limiting its ability to support fish and wildlife or human uses like swimming and safe drinking water.

Background
This lesson uses the Watershed Game developed by the University of Minnesota. The Watershed Game helps students understand connections between land use, clean water, and their community.

There are two types of game boards. One centers around a lake and the other around a stream. Each game has five land use maps: farm, forest, residential, city, and undeveloped areas that can be viewed on a screen or printed out on 11"x17" paper. These five land uses can be used in small groups and then be fit together to form a complete watershed. Alternatively, if playing as a whole class, all of the land uses can be also viewed on one single map or on a screen.

Strengths Used

Materials
- The Watershed Game: Classroom Version takes advantage of the excitement and interest of a board game to teach about ways we can reduce water pollution in our community. The goal of the activity is to reduce water pollution from various land uses to the stream without bankrupting the town. Working in teams, students practice collaboration, leadership, making persuasive arguments, and of course, math and science to apply tools (i.e. practices, plans, and policies) to decrease water pollution while balancing finanial resources.
- Schools and individuals can either purchase a complete game kit by completing the order form found on the website (watershedgame.umn.edu), or download and print the game components. The downloadable version is on the "Teachers" section of the "Classroom Version" dropdown menu. Note: self-printing requires multiple full-color pages and varying sizes of paper.

Procedure
- This game can be played in groups or as a class. The kit or printed version provides a tactical way for participants to play and an optional electronic scorecard helps the instructor and players keep track of efforts towards pollution reduction.
- To play, follow the detailed instructions included with the game. Feel free to modify the game to make it work for your students. Advanced players can add juggling financial resources and younger or non-reading students can use this game like a matching puzzle.

The grey house has roof runoff going into the stream on the game board. The "tool card" has a raingarden to capture the runoff and stop the pollution.

The construction site doesn't have silt fencing to contain the erosion. The "tool card" has silt fencing to contain the erosion.

ACTIVITY 8 — 20-30 Minutes

MISSION 3 WARM-UP ACTIVITY

HOW IS U.S. FRESHWATER USED?

Summary and Purpose
Students guess how water is used and then compare the real answers with their guesses.

Background
It will likely come as a surprise how much water coal and nuclear power generation consume. As told in the story, according to the Union of Concerned Scientists, the U.S. uses almost three times the amount of water flowing over Niagara Falls every single minute to generate electricity!

"Take the average amount of water flowing over Niagara Falls in a minute. Now triple it. That's almost how much water power plants in the United States take in for cooling each minute, on average.

In 2005, the nation's thermoelectric power plants—which boil water to create steam, which in turn drives turbines to produce electricity—withdrew as much water as farms did, and more than four times as much as all U.S. residences.

It requires more water, on average, to generate the electricity that lights our rooms, powers our computers and TVs, and runs our household appliances, than the total amount of water we use in our homes for everyday tasks—washing dishes and clothes, showering, flushing toilets, and watering lawns and gardens."

(Freshwater Use By U.S. Power Plants: Electricity's Thirst For a Precious Resource (2011) https://www.ucsusa.org/clean_energy/our-energy-choices/energy-and-water-use/freshwater-use-by-us-power-plants.html)

Water used in power generation is often classified as "non-consumptive" because the water is just moved from one place to another. For example, the water is pumped out of the ground or rivers, used in the power plant to create steam to run the equipment and then released back into rivers. However, the term "non-consumptive" is being challenged because the groundwater that is pumped up is no longer available to be used as drinking water in that region. In addition, the water that is released back into the rivers is hot (thermal pollution) and damaging to aquatic life.

Procedure
Brainstorm and Categorize Water Uses
Create a list (as a class or in groups) of the different ways water is used in *society*—not just in their homes. After a list is generated, consolidate the uses into the following categories.

    IRRIGATION—to grow food crops, lawn irrigation is under "public supply"
    AQUACULTURE—fish farming
    PUBLIC SUPPLY—all home/community uses, e.g. drinking water, cooking, toilets, pools, etc.
    DOMESTIC, "self-supplied" (or private well)—homes and community uses that have their own wells (this use has been combined with public supply on the pie chart)
    MINING—to mine minerals from the Earth to produce products
    INDUSTRIAL—to manufacture products (this use has been combined with mining on the pie chart)
    THERMOELECTRIC POWER—coal and nuclear power plants

Rank Water Uses
Using the handout on the page, have students rank the water uses. After the students find out the actual answers, have the students rank what ***they*** consider the most important three uses of water to be. This can be done individually, in groups, or as a class.

Creating a Vision for a Water-Wise World
Have students remember what they considered to be the most important uses of water. Have the students create a chart reflecting how they think a water use chart SHOULD look if they were to decide who got to use how much water for various uses. Have the students share their charts with others and compare and contrast how their chart is the same or different from their peers' and the actual use chart.

    Next, have the students identify if any changes would need to happen to fulfill the students' visions of a water wise world. As a class, walk through the top water use categories (power generation, food production (irrigation/aquaculture) and public supply and identify things that could be done differently to reduce water use. This would also make a great research project or opportunity to create a poster demonstrating these things.

Here are some suggested prompts to elicit a productive discussion.
POWER GENERATION—
- Could they reduce their electricity use? Can they think of specific ways? Check these websites for tips that may not have been thought of:
    - https://green.harvard.edu/tools-resources/poster/top-5-steps-reduce-your-energy-consumption
    - https://green.harvard.edu/action/student-guide
- Are there opportunities to support wind, solar or geothermal in your communities? Check out this website where the user puts in their zip code to find the nearby solar

gardens. https://www.cleanenergyprojectbuilder.org/solar-gardens
- Could they consume fewer products and/or not buy products of poor quality?
- Buy items second-hand instead of new?
- Eliminate one-time use products or products with excess packaging?

FOOD PRODUCTION—
- How does wasting food waste water?
- How could growing food in home or community gardens conserve water (and reduce carbon footprints)?
- Does what you eat impact how much water you use? (SEE ACTIVITY 5, p. 41-43)

PUBLIC SUPPLY—
- Judging from the information shown below the from the Water Use It Wisely website (https://wateruseitwisely.com/water-tower/), does the class think efforts should be focused indoors or outdoors to reduce water use? (Outdoors is largest use with "lowest hanging fruit" or easiest way to make an impact with little effort.)
- Are there ways people could landscape to use less water? (Definitely! Plant native flowers that do not require irrigation, plant "low-mow" or "no-mow" lawns that do not require watering—or mowing! These yards also benefit pollinators. For more information, visit bluethumb.org)
- What can be done inside the house? (Use low-flow showers, toilets, fix leaks, etc.)

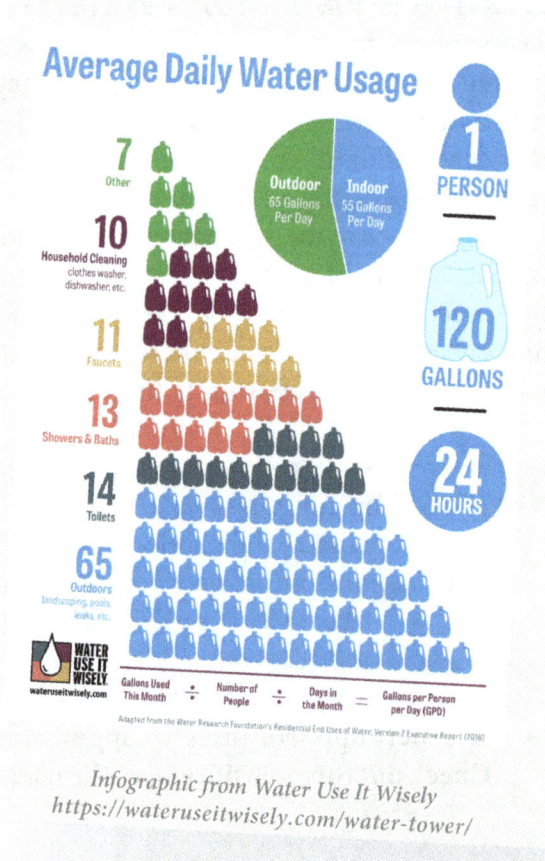

Infographic from Water Use It Wisely
https://wateruseitwisely.com/water-tower/

Please note that the best and highest use of water is for drinking. STUDENTS SHOULD BE REMINDED NOT TO DRINK LESS WATER IN ORDER TO CONSERVE WATER!

How do you think freshwater is used in the U.S.?

Please match the words from the word bank with the slices of "pie." It's OK to guess!

Billion of Gallons of Water Per Day

- **aquaculture** (fish farming)
- **irrigation** (watering crops)
- **livestock** (farm animals)
- **mining/industry**
- **public supply/private wells** (for homes and in the community)
- **thermoelectric power** (electricity generation from coal and nuclear)

Pie chart values: 133, 118, 42.3, 18.8, 7.6, 2

Answers — Does the reality of our country's water use match your ideal vision for water use? Please write your answer on the back.

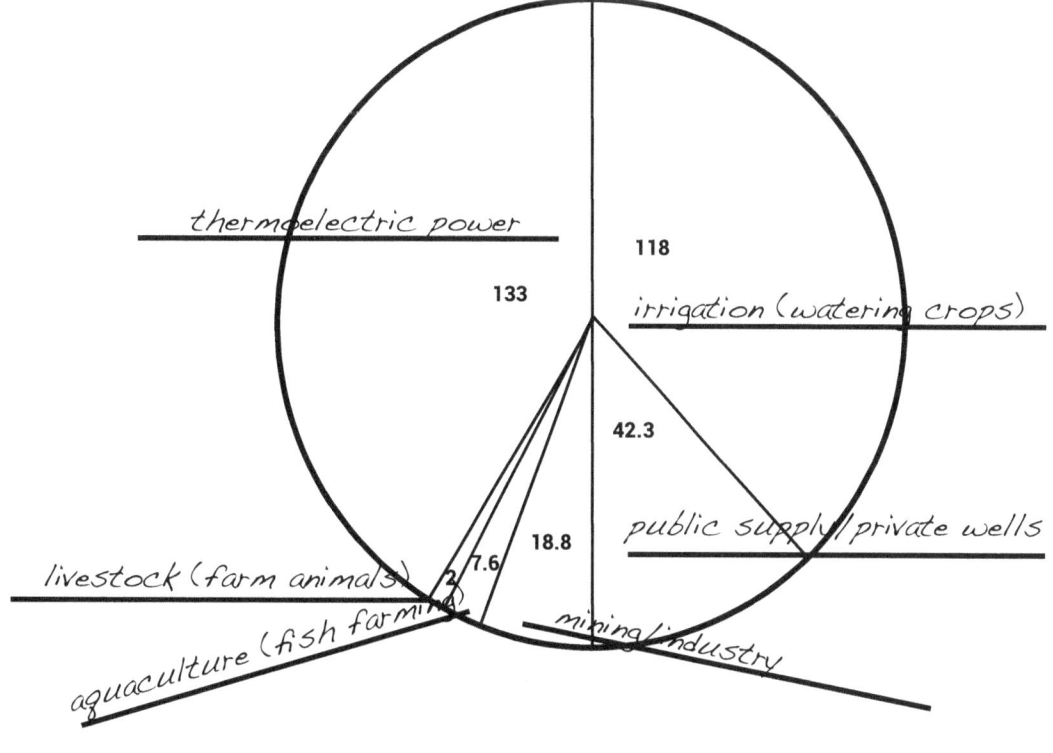

Information from USGS website: https://www.usgs.gov/media/images/2015-water-use-withdrawals-category

MISSION 3

Water	"Alrighty, then," agreed Water. **"MISSION THREE: USE LESS ENERGY!"**
Narrator	The kids' enthusiasm evaporated as they stared at Water with scrunched up, tilted faces and mouths **agape**. They were now thinking Water was a little crazy. Finally, a brave boy asked, "Why save energy? I thought we were protecting *WATER!*" Of course, wise old Water expected this reaction. Water wore a **solemn** expression and bobbed up and down as if saying 'yes.' Then Water sighed, took a deep breath, and calmly explained,
Water	"It's hard for humans to see that creating energy interrupts my water cycle in two significant ways. First, the burning of **fossil fuels** (e.g. coal, oil, and natural gas) is causing our planet to warm. A warmer planet means warmer air and more evaporation. And since warmer air can hold more water, more intense storms are becoming the new normal. Some people are calling these intense storms "rain bombs." Next, creating energy from coal, natural gas, and nuclear power plants takes an ENORMOUS amount of water—like billions of gallons a day—out of the ground and out of rivers. It's like the water cycle sprang enormous leaks!"

MISSION 3 CONTINUED

Water Water could see the kids were confused, so Water explained, "Imagine **THREE TIMES** the amount of water flowing over Niagara Falls every single minute. That's how much water the U.S. uses to generate the power we use to heat and cool our homes, turn on our lights and electronics, and everything else."

Kid 3 "DUDES," observed an older defender, "Look how tiny that boat looks compared to that waterfall! That's a craaa-zy amount of water."

All Kids **"Oh no! That's a LOT of water,"** hollered the heroes.

Water "Yes it is. Perhaps even more astonishing," explained Water, "is that it requires more water to make the electricity for homes than the total amount of water we use in our homes for everyday tasks like washing dishes and clothes, showering, flushing toilets, and watering lawns and gardens."[1]

---

[1] Freshwater Use By U.S. Power Plants: Electricity's Thirst For a Precious Resource (2011) https://www.ucsusa.org/clean_energy/our-energy-choices/energy-and-water-use/freshwater-use-by-us-power-plants.html

MISSION 3 CONTINUED

Kid 4   "Wait, but isn't the water used by power plants just put back into the river?" questioned a quizzical kid.

Water   "Very clever reasoning," said Water. "I'm so glad you brought this up. It's true that many people label water for energy production as "non-consumptive"—meaning the water isn't really being consumed or used. These people argue that the water is just "borrowed." But the thing that people don't think about is that the water that returns to the river is much too hot for the aquatic life—like plants, fish and **macro-invertebrates**. Some like it hot, but trout do not.

Water   Not to mention the fish don't appreciate the mercury that gets released into the atmosphere when burning coal and rains down into our lakes and streams. Finally, a lot of the water used in power plants is taken out of the ground. That means it's no longer available for the people who live there to use as drinking water.

All Kids   **"OK! We accept that mission to save energy, Water!"**

Narrator   The kids made a list of what they could do and what they could talk to their adults about doing. The list included small steps like shutting off lights. It also included bigger steps like supporting wind and solar. As they kept thinking, they decided they wanted to go big and change policy to encourage energy-efficient building design techniques—like passive design that use 90% less energy!

MISSION 3

HOME ENERGY USE

SUMMARY AND PURPOSE
Learn about the "passive home"—which uses 90% less energy simply by how it is built. Have students continue to brainstorm ways they can reduce their energy use at home.

BACKGROUND
When we think about energy use impacts, we typically think about climate change, not water conservation. The fact that our energy use has huge negative impacts on both water and climate underscores the urgent need for a lot more attention and education about energy use. While small changes like turning off lights are important, system changes that make larger impacts are needed.

In the Midwest, over half of our home energy use is for heating and cooling our homes.* What if we could virtually eliminate this energy need by building differently? Good news! We can by using innovative building techniques. The passive house is one example of this. This 90 second video (Passive House

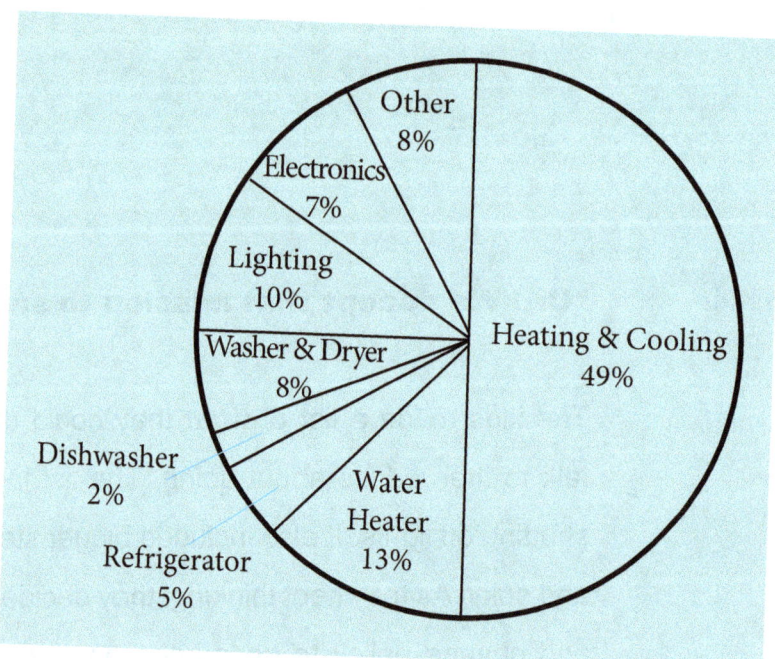

Information from NASA Climate Kids
https://climatekids.nasa.gov

Explained in 90 Seconds by Hans-Jörn Eich https://vimeo.com/74294955) gives a nice introduction to passive housing. To find out more about passive housing (or schedule a class tour), visit https://passivehouseminnesota.org/.

> * In Minnesota, most furnaces use natural gas. Hydraulic fracturing, or "fracking," is a drilling process that injects millions of gallons of water, sand and undisclosed chemicals at high pressure into horizontal wells to crack open shale rock and release natural gas. While the natural gas has different issues associated with it than are addressed in this book, the impacts on water are still unfavorable.

STRENGTHS USED

MATERIALS
Smart board/Computer projector to show 90-second video

PROCEDURES
Show students the pie graph on page 36 and ask them where they think it would make sense to tackle reducing energy use. (There is clearly no wrong answer here!) When the suggestion to tackle home heating and cooling comes up, show the 90-second video (Passive House Explained in 90 Seconds by Hans-Jörn Eich https://vimeo.com/74294955) to introduce a whole different way of approaching building.

Follow up discussion ideas:
1. Why isn't more of this being done now? (It's relatively new and it costs more to build this way. However, it pays for itself quickly and has long-term positive environmental benefits.)
2. Why don't people think long term?
3. What kind of house will you live in? How will you get to work?

MISSION 4

All Kids	"What else can we do?" asked another water defender.
Water	Water smiled and whispered, **"MISSION FOUR: DON'T WASTE FOOD."**
Narrator	Again, the kids were bewildered. What did wasting food have to do with helping water? Water predicted the

confusion and quickly turned into a giant irrigation system! The kids shrieked as Water soaked them. The kids needed no further explanation. They experienced firsthand that producing food requires enormous amounts of water too.

MISSION 4

Narrator   The defenders of the future looked around and at their feet. They weren't so sure they wanted to commit to not wasting food. After all, their parents had a history of serving healthy food that didn't always taste very good.

Kid 4　　Finally the blonde **bellered**, "We can do that!" With that exclamation, the other kids realized that they could—and should—do that too. This was an important issue and sometimes it takes a little personal sacrifice.

Water　　"Thanks, kids! Now you know me. Water never stays in one place, so I need to get moving now."

Narrator　　The defenders said goodbye to Water and thanked Water for the important lessons. Water turned and smiled. Are you wondering what Water said back?

CONCLUSION

Narrator   Nothing.

                   Water just waved.

ACTIVITY 10 — 45-60 Minutes

MISSION 4

HOW MUCH WATER DO YOU EAT?

Summary and Purpose
It's important to think about the entire "life" the food and products we consume to consider ways to minimize our impact on the world.

Background
As award-winning novelist, poet, essayist, environmental activist, cultural critic, and farmer, Wendell Berry said, "Eating is inescapably an agricultural act, and how we eat determines, to a considerable extent, how the world is used." According to a May 2018 study by Proceedings of the National Academy of Sciences of the United States of America, 60% of the world's mammals are livestock, 36% are humans, and only 4% are wildlife. That means 96% of the world's mammals are either people or food for people. What people eat has huge implications for greenhouse gas emissions, biodiversity and also water use.

STRENGTHS Used

Materials
- Paper
- Pencils
- Handout or information from page 39 presented on a screen.

Procedures

Have students write down what they typically eat for breakfast, lunch and dinner.

Next, have the students add up the water they eat (indirectly) on a typical day using the chart below.

FOOD ITEM	SERVING SIZE	WATER FOOTPRINT
Steak (beef)	6 ounces	674 gallons
Hamburger	1 (includes bread, meat, lettuce, tomato)	660 gallons
Ham (pork)	3 ounces	135 gallons
Eggs	1 egg	52 gallons
Soda	17 ounces	46 gallons
Coffee	1 cup	34 gallons
Salad	1 (includes tomato, lettuce, cucumbers)	21 gallons

Data from the Water Footprint Network

For more detailed information, check out these resources:

https://www.youtube.com/watch?v=4Bu0nWY_lYw
https://thewaterweeat.com/

Can they explain why meats use more water to make than plant crops? (This is an opportunity to briefly bring in the concept of primary and secondary producers.) Estimate how much water it takes to make each meal. About how much water do they need to eat for a week? A month? A year? A decade? A lifetime? After reviewing the information about how much water it takes to produce different types of foods, would they consider making any dietary changes? **(NOTE: Again, please stress that the best use of water is for drinking. Discourage any discussion about drinking less water!)**

If time allows, visit https://waterfootprint.org/ to have students take more in-depth water inventories. There are also resources such as slide presentations for schools about water footprint: https://waterfootprint.org/media/downloads/WFN_presentation_schools.pdf

How much water do you EAT?

Breakfast	gallons	liters	Snacks/Treats	gallons	liters	Lunch	gallons	liters	Dinner	gallons	liters
Egg	36	135	Chocolate	450	1703	Bread (1 slice)	11	40	Steak (8 oz.)	850	3,218
Toast (1 slice) with butter add 14 gallons	11	40	Chips	18	68	Hummus	10	38	Hamburger (includes bread, meat, lettuce, tomato)	660	2,498
Bacon (2 pieces)	202	765	Cookie	20	76	Peanut butter and jelly/Nutella	3	11	Sausage/hotdog	464	1,755
Pancakes	12	45	Cake	30	114	Applesauce (in plastic container)	53	201	Ham/Pork (3 oz.)	135	511
Yogurt	35	132	Candy (12 Starbursts)	29	110	Cheese (1 slice)	25	500	Chicken	81	307
Greek yogurt	90	341	Crackers	11	40	Fruit			Veggies		
Fruit (see lunch)		70				Oranges	28	106	Tomatoes	11	42
						Apples	42	159	Cabbage	12	45
			Hummus	10	38	Bananas	51	193	Cucumber	14	53
						Peaches	71	269	Corn	54	204
						Mango	95	360	Lettuce	8	30
									Potatoes	18	68
Milk (dairy)	30	114	Ice cream	42	159	Tap water	0.12	0.45	Cauliflower	20	114
Milk (soy)	9	34				Bottled water/pop	33	125	Tofu	81	307
Milk (almond)	23	87							Beans/Lentils	10	38
Juice (apple)	75	284							Pizza (cheese)	49	185
Coffee	38	144							Pasta/Spaghetti	12	45
Tea	10	38							Rice	98	371
Cereal	20	76							Macaroni & cheese	40	151

PRODUCTION

Water Footprint measures how much freshwater is used from beginning to end of a product

What do you usually eat for...

Breakfast	Snack	Lunch	Dinner

Total ____   Total ____   Total ____   Total ____

Daily Total ____

Introduction video found at
https://www.youtube.com/watch?v=4Bu0nWY_1Yw

ADDITIONAL WATER RESOURCES...

Blue Thumb—Planting for Clean Water
bluethumb.org

Everything you need to get started on planting native plants.

Master Water Stewards
https://masterwaterstewards.org/

Freshwater Society's Master Water Stewards program provides opportunities for residents to develop expertise in improving and protecting water resources through training, hands-on projects, and outreach. Following certification, the Master Water Stewards can help people in the watershed who are looking for ways to improve their local water resources through specific outreach and action.

Metro Area Children's Water Festival
http://metrocwf.org/

The Metro Area Children's Water Festival is an annual event for fourth graders that is held the last Wednesday in September at the Minnesota State Fairgrounds. The festival teaches students about water resources and ways to protect water and are staffed by environmental experts from water agencies and organizations. All students attend the Science Museum of Minnesota presentation and enjoy the Water Arcade.

Teachers need to register for the opportunity to participate and schools are selected by lottery. The Children's Water Festival website contains the registration form, plus information on how to sponsor or volunteer.

St. Croix River Association, Rivers are Alive Program
https://www.stcroixriverassociation.org/river-connections/

The Rivers Are Alive K-12 Environmental Education Program is a National Park Service program that has served the St. Croix River watershed for more than 20 years. SCRA is a key partner in helping to coordinate and facilitate Rivers Are Alive classroom presentations and field trips. Rivers Are Alive offers immersive, nature-based, hands-on, and standards-aligned activities that inspire young people to connect with nature and become stewards of our wild places. Pre and post activities are also available to deepen the experience for your students.

Salt Education
SaltSmart.info

Keep streets and sidewalks safe AND protect water.

Taking Aim at The Plastic Plague
This 60 Minutes episode is an eye-opening documentary about how pervasive and problematic plastics are in our oceans and how a young man, Boyan Slat, is working to solve the problem. After viewing this video, kids may have a different perspective about plastic.

U.S. Environmental Protection Agency (EPA)
Subscribe to EPA's Environmental Education eNewsletter to receive monthly e-news to help educators connect with federal programs and new resources.

free download

Water Ways: A Minnesota Water Primer & Project WET Companion

Table of Contents
Chapter 1: Water Basics
Chapter 2: Minnesota Waters: Atmosphere, Rivers, Lakes
Chapter 3: Minnesota Waters: Wetlands and Groundwater
Chapter 4: Life in Water
Chapter 5: Using Water
Chapter 6: Harm and Hope
Chapter 7: Governing Water, A Minnesota Water Timeline

Appendices
Appendix 1: Minnesota Watershed Maps
Appendix 2: Sample Minnesota Water Case Study: 100th Anniversary of Safe Drinking Water for Minneapolis
Appendix 3: Contacts for Local and State Water Information
Appendix 4: Selected Sources and Websites
Glossary
Index

Additional Resources
Career profiles
Classroom Connections (activity suggestions)
Out and About (field trip suggestions)

GLOSSARY

The terms are in bold in the book text and they are listed below in they order they appear in the story and lessons.

Introduction

mesmerized (p. 13) to mesmerize is to hold someone's complete attention. Synonyms: enthrall, spellbind, entrance, hold spellbound, dazzle, bewitch, charm, captivate, enrapture, enchant, fascinate, transfix, hypnotize

Anthropocene (p. 14) "*anthropo*" is greek for "human" and "*-cene*" is used for geologic time periods. The time period we are living in is sometimes referred to as "Anthropocene" because humans are shaping the period.

nucleus (p. 15) the core or central part of something (like a cell)

oxygen-rich p. 15) containing a lot of oxygen

atmosphere (p. 15) the gases surrounding the Earth

terrestrial (p. 15) land

biocrusts (p. 15) short for biological soil crusts, biocrusts are like a "living skin" at the soil surface communities comprised of cyanobacteria, mosses, liverworts, fungi, algae and lichens

cyanobacteria (p. 15), bacteria that get their energy through photosynthesis. The name cyanobacteria comes from the blue color of the bacteria.

multi-cellular organisms (p. 15) are organisms that consist of more than one cell

precipitation (p. 16) all of the forms of water particles (whether liquid or solid) that fall from the atmosphere (e.g., rain, hail, snow or sleet)

infiltration (p. 16) soak in, percolate

collection (p. 16) a place where water sits, such as a lake or river

evaporation (p. 16) evaporation of water occurs when the surface of the liquid is exposed, allowing molecules to escape and form water vapor; this vapor can then rise up and form clouds. With sufficient energy, the liquid will turn into vapor.

condensation (p. 16) the change of matter from the gas phase into the liquid phase. Condensation is the reverse of evaporation.

Mission 1: Let the Water Soak Into the Ground

recharge (p. 30) is where water moves downward from the Earth's surface into the ground. (Recharge is the primary method through which water enters an aquifer.)

impervious (p. 31) not allowing fluid to pass through

raingarden (p. 32) a shallow depression, or hole, planted with trees, shrubs and/or flowers designed to temporarily hold and soak in rain water runoff that flows from roofs, driveways, patios or lawns.

pervious (p. 33) allowing fluid to pass through

watershed (p. 35) is an area of land that drains into a particular water body.

stormwater runoff (p. 40) is rainfall that flows over the ground. It is created when rain falls on roads, driveways, parking lots, rooftops and other paved surfaces that do not allow water to soak into the ground

best mangaement practices (BMPs) (p. 41) are methods or techniques found to be the most effective and practical means in achieving an objective (such as preventing or minimizing pollution) while making the optimum use of resources

stormdrain (p. 42) is a drain built to carry away excess water in times of heavy rain

Mission 3: Save Energy

thermoelectric power generation (p. 46) Electric power generated from a heat source, such as burning fossil fuel-coal, oil, indirectly through devices like steam turbines.

irrigation (p. 47) watering (typically lawns or crops)

public (water) supply (p. 47) water that is used by the public for homes, places of business and in the community.

private (water) wells (p. 47) water pumped up from a well on a landowner's property

aquaculture (p. 47) rearing of aquatic animals or the cultivation of aquatic plants for food

livestock (p. 47) farm animals

agape (p. 50) wide open, especially with surprise or wonder

fossil fuels (p. 50) Fossil fuel is a general term for organic materials formed from decayed plants and animals from hundreds of millions of years ago. The three types of fossil fuels commonly used to make energy are coal, oil and natural gas.

solemn (p. 50) not cheerful or smiling; serious.

macro-invertabrates (p. 54) are organisms that lack a spine and are large enough to be seen with the naked eye. Examples of macro- invertebrates include flatworms, crayfish, snails, clams and insects, such as dragonflies

Mission 4: Save Food

bellered (p. 61) holler or yell

www.ingramcontent.com/pod-product-compliance
Lightning Source LLC
Chambersburg PA
CBHW060427010526
44118CB00017B/2387